6th Grade Math Volume 1

Newburyport, MA 01950
800-596-3175
www.onboardacademics.com

ISBN: 978-1494857271

Table of Contents

Order of Operation

Key Vocabulary

order of operations

parentheses

simplify

Order of Operations
Study the chart to learn the order of operations.

Left to right →

①	**Parentheses**	
②	**Multiplication**	**Division**
③	**Addition**	**Subtraction**

Perform calculations inside parentheses

From left to right, perform multiplications and divisions

From left to right, perform additions and subtractions

---

4 + 4 x 3
Place a check mark next to the student who answered correctly. Use your order of operation rules.

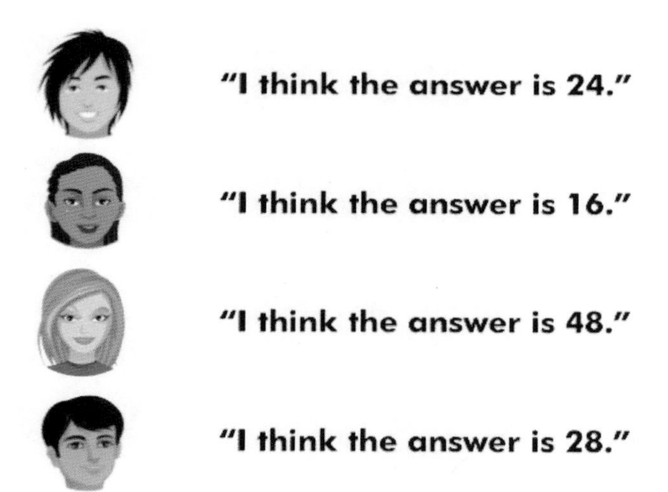

Who is right?

Order these cards to express the correct rules for order of operations.
Write the operation in the correct box.

First

Last

Division
Addition
Parentheses
Multiplication
Subtraction

Complete a problem.

Place a check mark to indicate the correct answer. Show your work next to the order of operation listed below.

$30 - 6 \cdot (5 + 3) \div 2 + 8 =$

104	6	14	19.2

Parentheses

Multiplication

Division

Subtraction

Addition

Solve these order of operation problems.

$8 \div 2 + 8 - 6 =$ ☐

$2 + 3 \times 4 - 5 =$ ☐

$2 \times 3 + 5 - 2 =$ ☐

$24 \div 3 \times 8 + 4 \times 3 =$ ☐

Order of operations challenge.

Write in the parentheses in the proper location for each equation and then solve.

( )

$3 \times 6 + 2 - 4 \times 5 + 1 =$ ☐

$3 \times 6 + 2 - 4 \times 5 + 1 =$ ☐

$3 \times 6 + 2 - 4 \times 5 + 1 =$ ☐

$3 \times 6 + 2 - 4 \times 5 + 1 =$ ☐

Name______________________________

Order of Operations Quiz

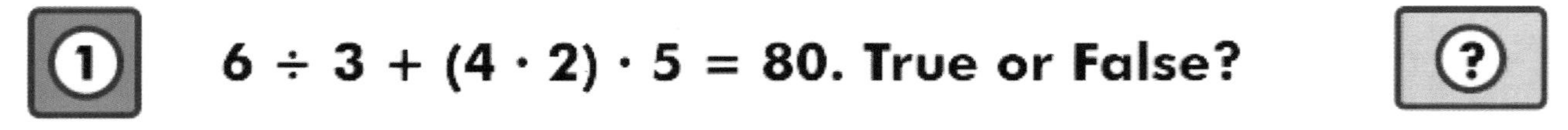

2. $3 + (3 \cdot 2) \cdot (4 - 2) = ?$

   A. 36

   B. 18

   C. 15

   D. 25

3. $4 \cdot 3 \cdot 2 + 2 + (4 + 2) = ?$

4. $24 - 3 \times (2 \times 3) = ?$

Multiples & Least Common Multiples (LCM)

Key Vocabulary

factor

multiple

least common multiple (LCM)

Multiple or Factor?

The multiple of a number is the result of multiplying that number by another whole number, e.g. 10 is a multiple of 5.

Write the correct description in the box provided.

16 is a		of 32
15 is a		of 5
8 is a		of 2
2 is a		of 4

Multiple

or

Factor

Multiples and Common Multiples

The first five multiples of 8:

The first six multiples of 4:

The first three multiples of 12:

Which number is common to all three lists?

We call this number a common Multiple. This number is also the least common multiple of 8, 4 and 12. This is sometime referred to by its initials LCM.

Find the multiples of 6 and 4 by drawing a circle around the multiple.

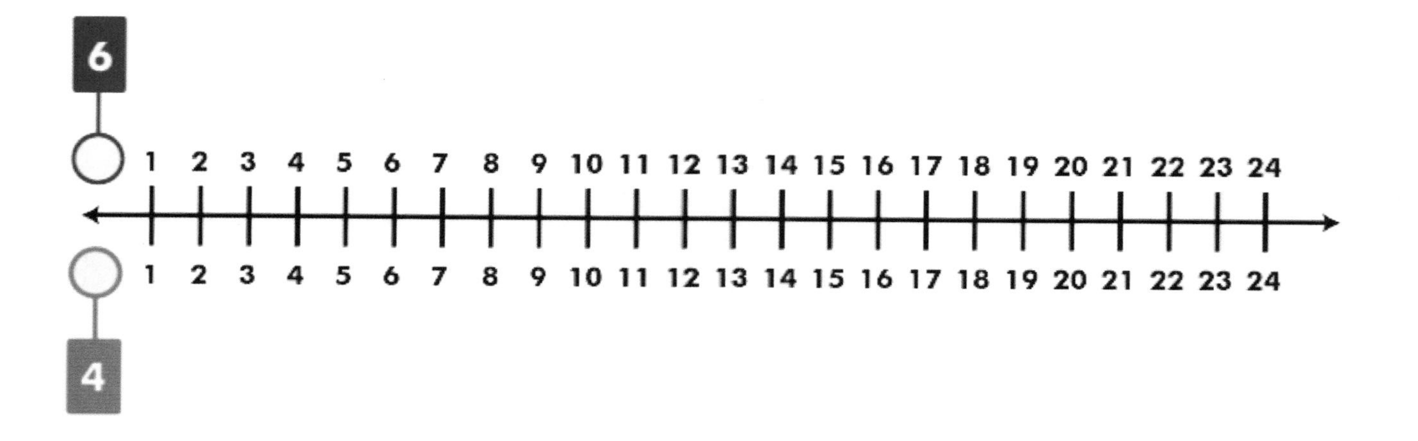

What are the common multiples of 4 and 6?

What is the least common multiple (LCM) of 4 and 6?

Using LCM to solve problems.
Draw a circle on the dates that Sanjeev goest to the gym and the the days that Amy goes to the gym. Use different colors for each student. Draw a green box on the date that they will meet at the gym.

Sanjeev goes to the gym every five days.

1	2	3	4	5	6	7
8	9	10	11	12	13	14
15	16	17	18	19	20	21
22	23	24	25	26	27	28
29	30	31				

Amy goes to the gym every four days.

When will Sanjeev and Amy meet at the gym?

When will they meet again?

Use the same method for identifying dates but start with their last meeting.

January

1	2	3	4	5	6	7
8	9	10	11	12	13	14
15	16	17	18	19	20	21
22	23	24	25	26	27	28
29	30	31				

February

			1	2	3	4
5	6	7	8	9	10	11
12	13	14	15	16	17	18
19	20	21	22	23	24	25
26	27	28				

Alarm and buzzer challenge.

The school's fire alarm is being tested and has been set to ring every 40 minutes throughout the day. The lesson buzzer sounds every hour. The school day starts at 8 AM and ends at 2 PM.

At what times during the school day will both the alarm and the lesson buzzer sound? Organize the images below to help you.

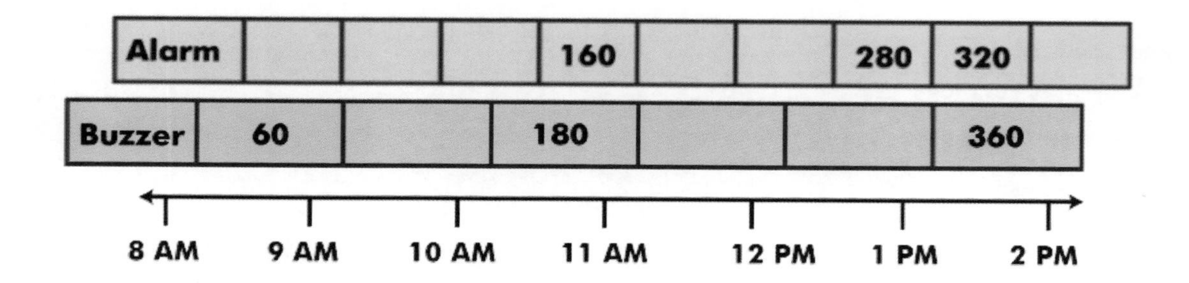

Name________________________________

Multiples & Least Common Multiples (LCM) Quiz

1. 16 is a multiple of 4. True or False?

2. What is the LCM of 5 and 10?

   A. 5
   B. 10
   C. 50
   D. 15

3. What is the LCM of 3 and 11?

4. What is the LCM of 6, 7 and 14?

Greatest Common Factor

Key Vocabulary

factor

greatest common factor (GCF)

Find the factors of 24 and 32.
If the factor is just for 24 write it in the yellow space, if its just for 32 write it in the blue space and if the factor is for both 24 and 32 write it in the green space.

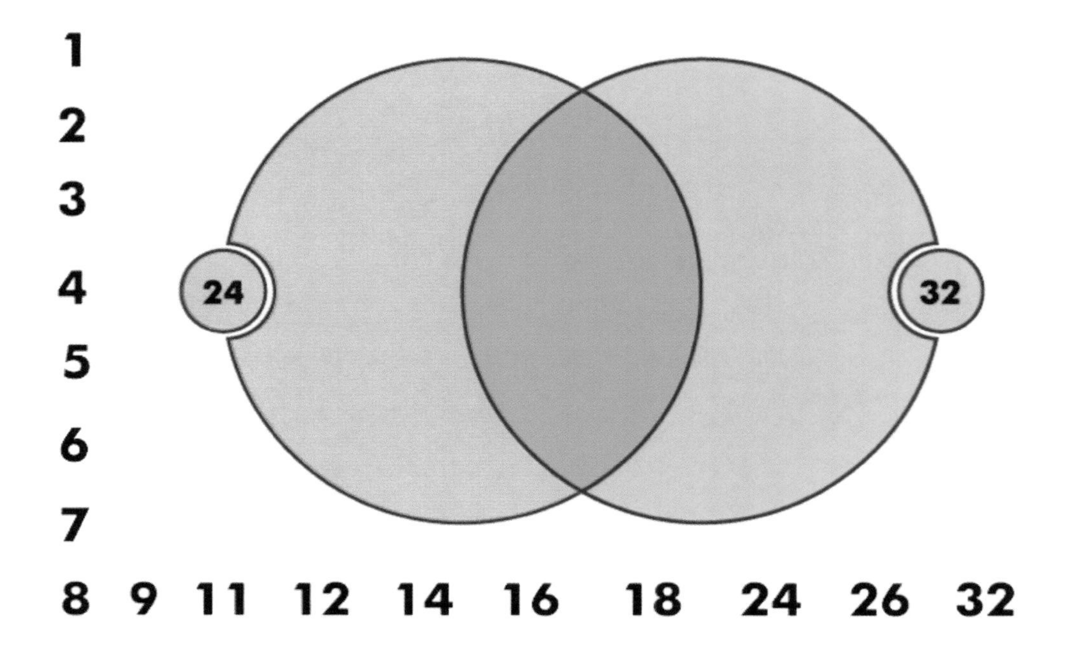

The numbers in the green space are the common factors..

What is the greatest common factor (the largest)? Circle it.

Finding the GCP of 30 and 10.

1 Find the factors of 30 | 1 | 2 | 3 | 5 | 6 | 10 | 15 | 30 |

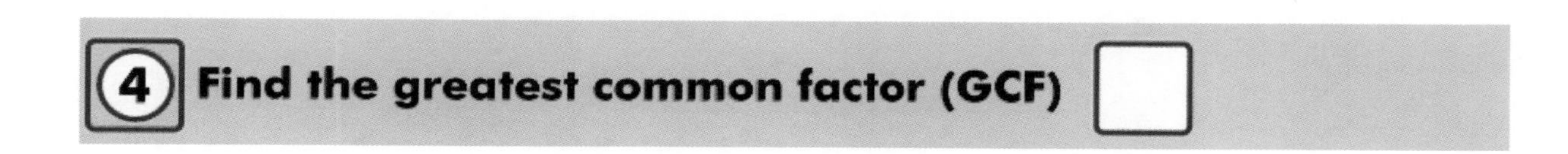

Finding the GCF using prime factors.

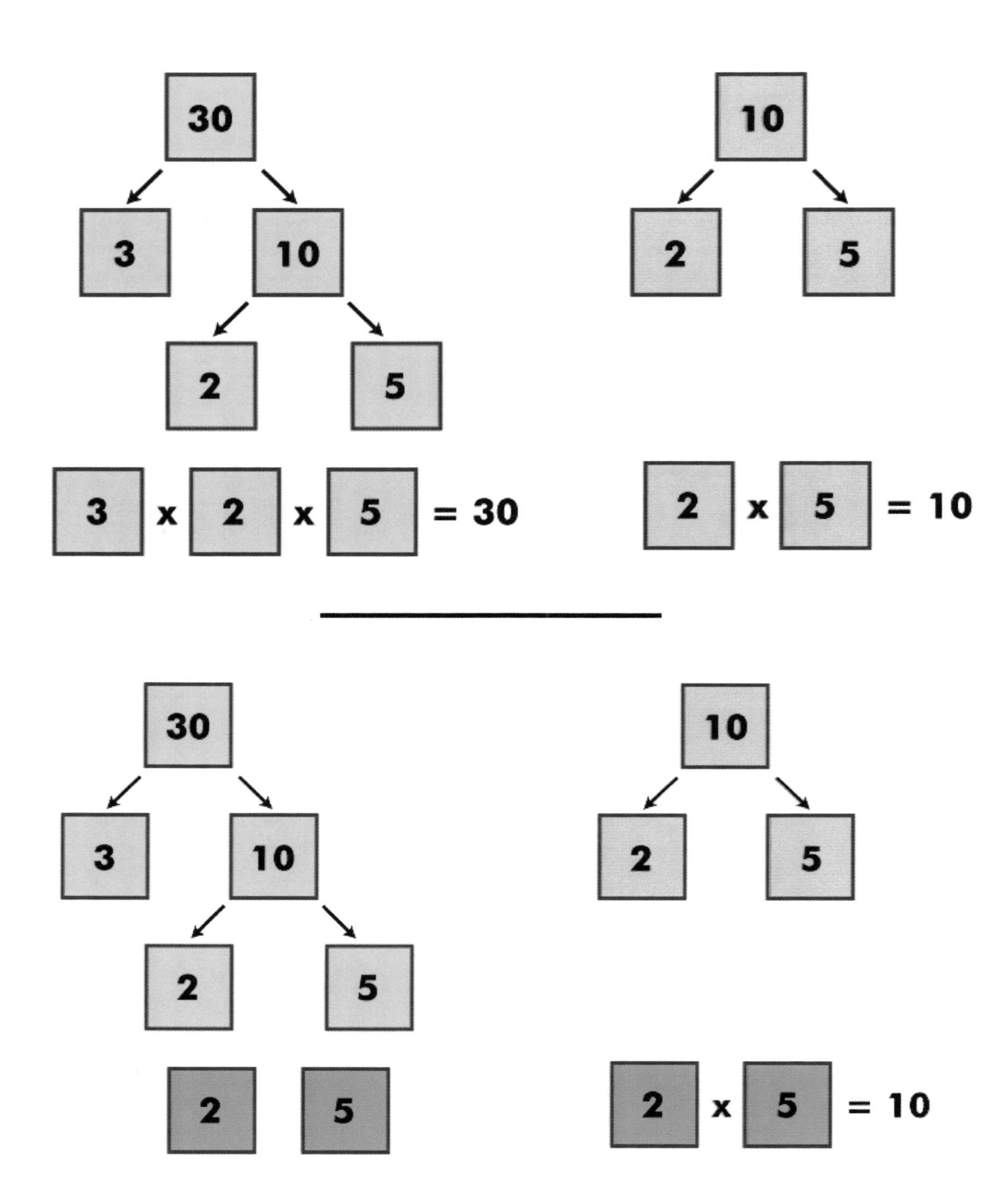

The GCF is the product of the common prime factors

Use the prime factors to find the GCF of 18 and 24.

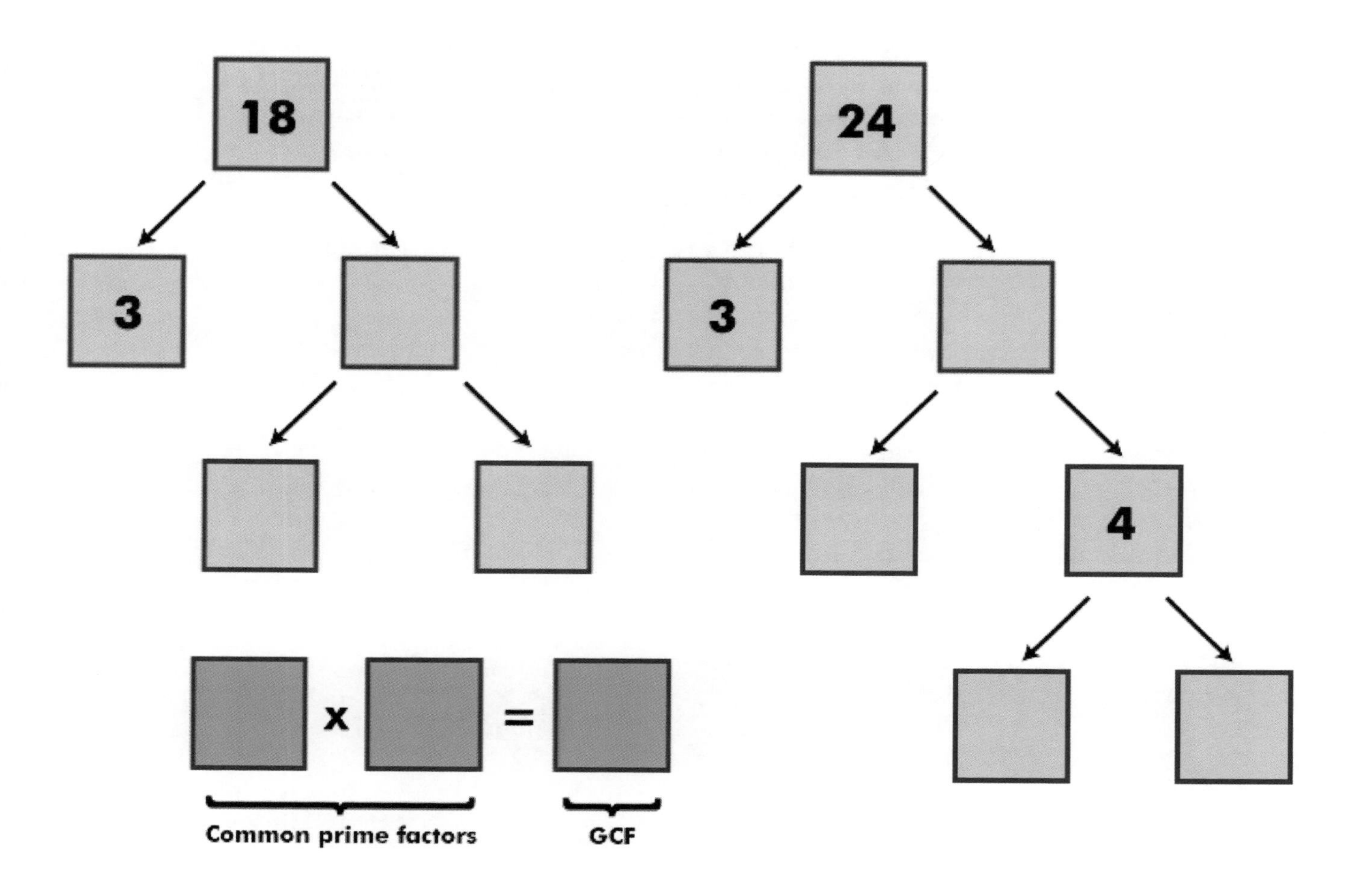

Solve this word problem using prime factors.

Sanjeev has a part-time job stacking shelves. His boss asks him to stack 42 boxes of *Iron-Bran* and 56 boxes of *Fruity Fiber*. He must stack them in such a way that each row has the same number of boxes, but he can not mix boxes in the same row.

What is the greatest number of boxes that can be in each row?

Name________________________________

Greatest Common Factor Quiz

1. The GCF of 8 and 16 is 4. True or False?

2. The common prime factors of 8 and 16 are:

   A. 2 x 2 x 2
   B. 2 x 2
   C. 2 x 4
   D. 2 x 2 x 2 x 2

3. What is the GCF of 9 and 42?

4. What is the GCF of 12, 16 and 24?

Prime Factorization

Key Vocabulary

prime

composite

prime factorization

Study this factor tree.

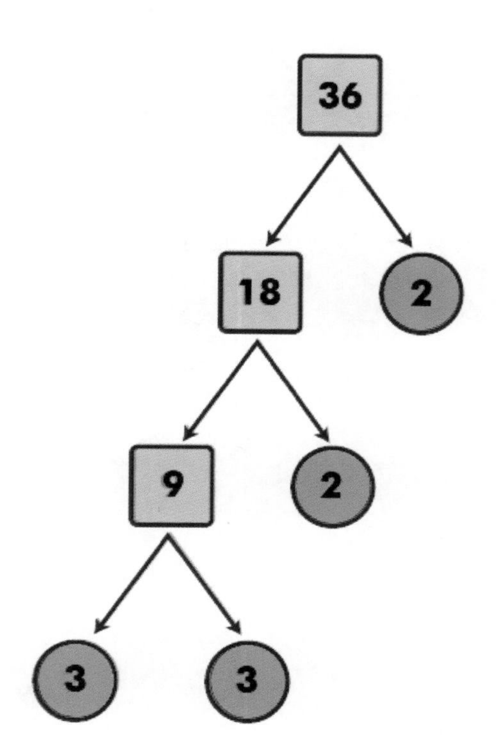

Solution

$$36 = 2 \times 2 \times 3 \times 3$$
$$= 2^2 \times 3^2$$

Key

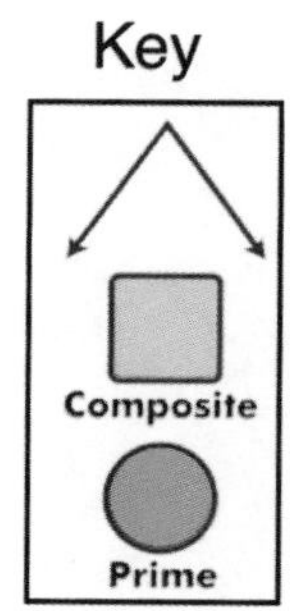

Complete the factor tree using the elements below.
There are two possible solutions.

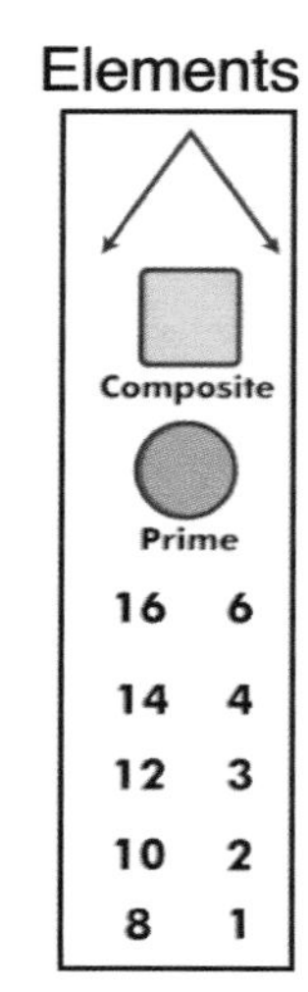

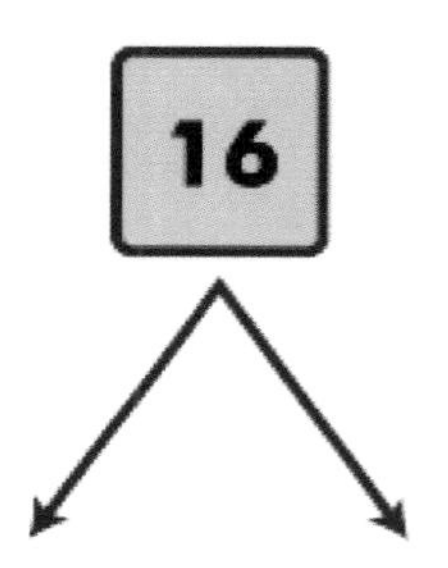

Complete the factor tree and write the prime factorization in exponential form.

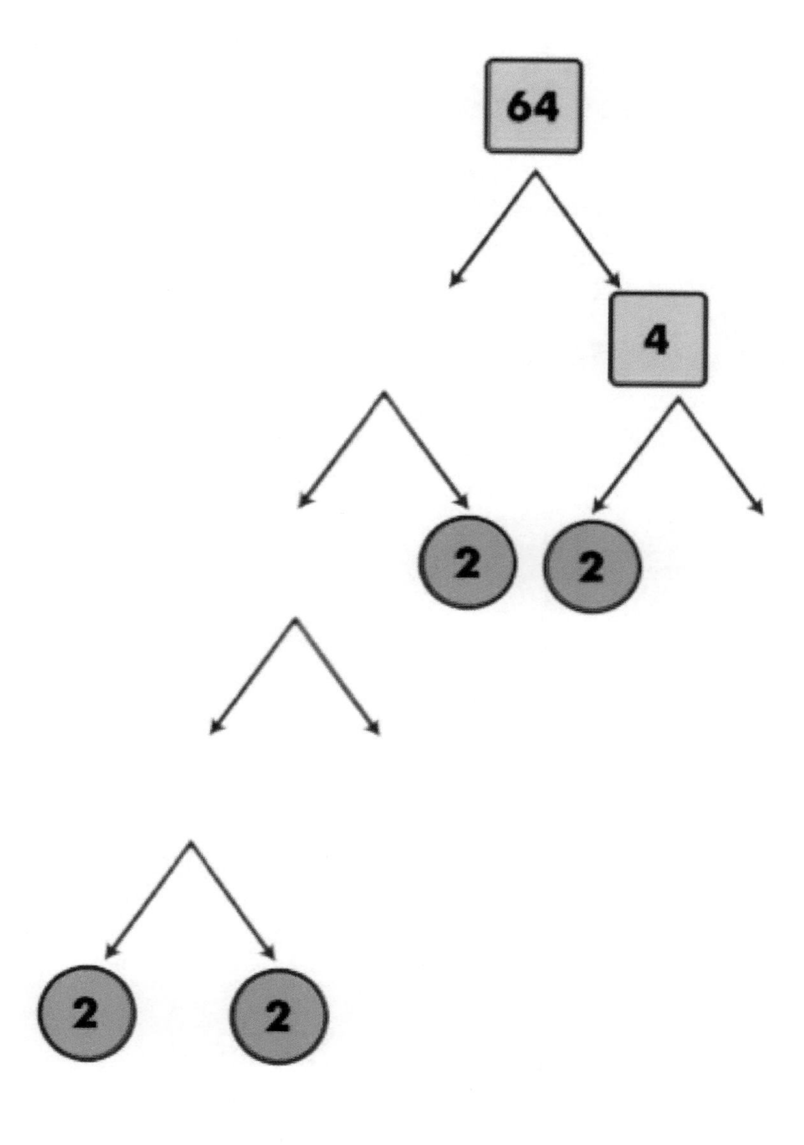

Name________________________________

Prime Factorization Quiz

1 True or false: the prime factorization of 32 is 2^6.

2 Which is the correct prime factorization of 36?

- A 3 x 3 x 4
- B 2 x 2 x 2 x 3 x 3
- C 2 x 2 x 3 x 3
- D 2 x 2 x 9

3 Find the value of y to complete the prime factorization: $2^y \cdot y^y = 216$

4 The prime factorization of which number is $2 \times 3^2 \times 7 \times 13$?

Compare & Order Fractions

Key Vocabulary

common denominator

equivalent fractions

Each friend has completed a portion of the charity bike ride.
Shade the boxes to represent the amount that each friend completed and then express it in 15ths.

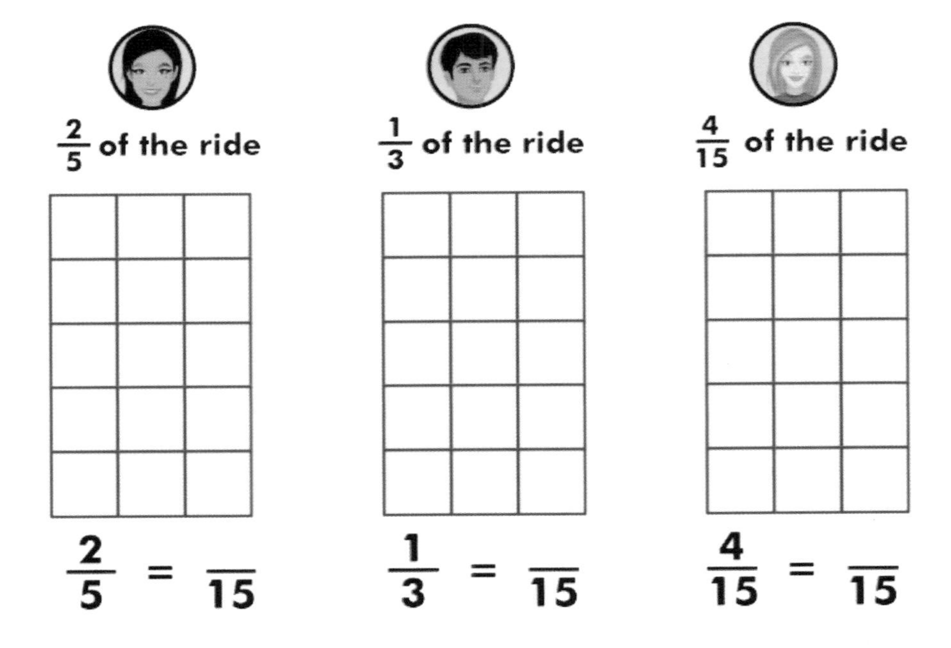

Who is closest to finishing?
Place a check mark next to the friend closest to the finish line.

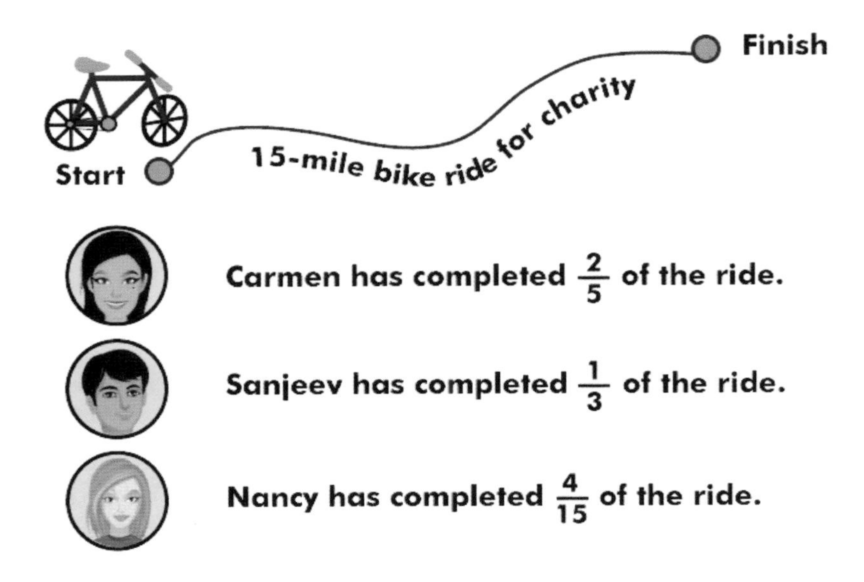

Which is larger 5/8 or 3/4?

What is the least common denominator (LCD)?

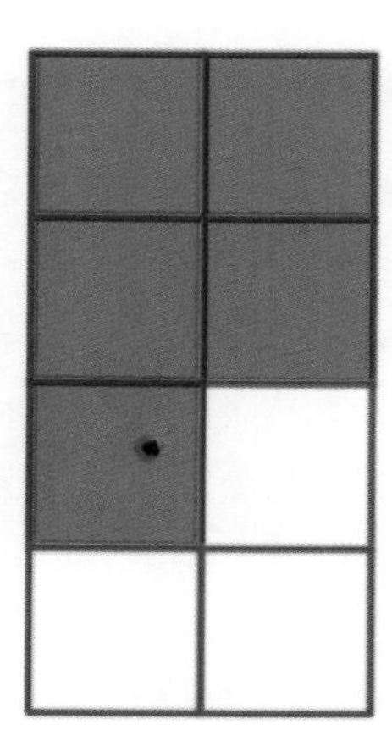

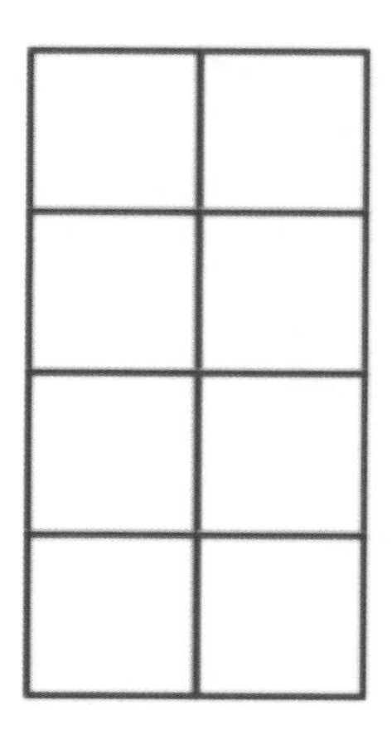

$\frac{5}{8}$

$\frac{3}{4} = \frac{\quad}{\quad}$

Compare Fractions

Rewrite the fractions using the least common denominator and then use one of the comparative symbols to complete the equation.

$\frac{2}{7}$ ◯ $\frac{3}{5}$ $\frac{7}{8}$ ◯ $\frac{3}{4}$

$=$

___ ___ ___ ___

$<$

$\frac{4}{5}$ ◯ $\frac{5}{6}$ $>$ $\frac{5}{10}$ ◯ $\frac{6}{12}$

___ ___ ___ ___

Who ate the most cookies?Here

When her friends came over to visit, Ashima's mother put out a plate of cookies. Who ate the most?

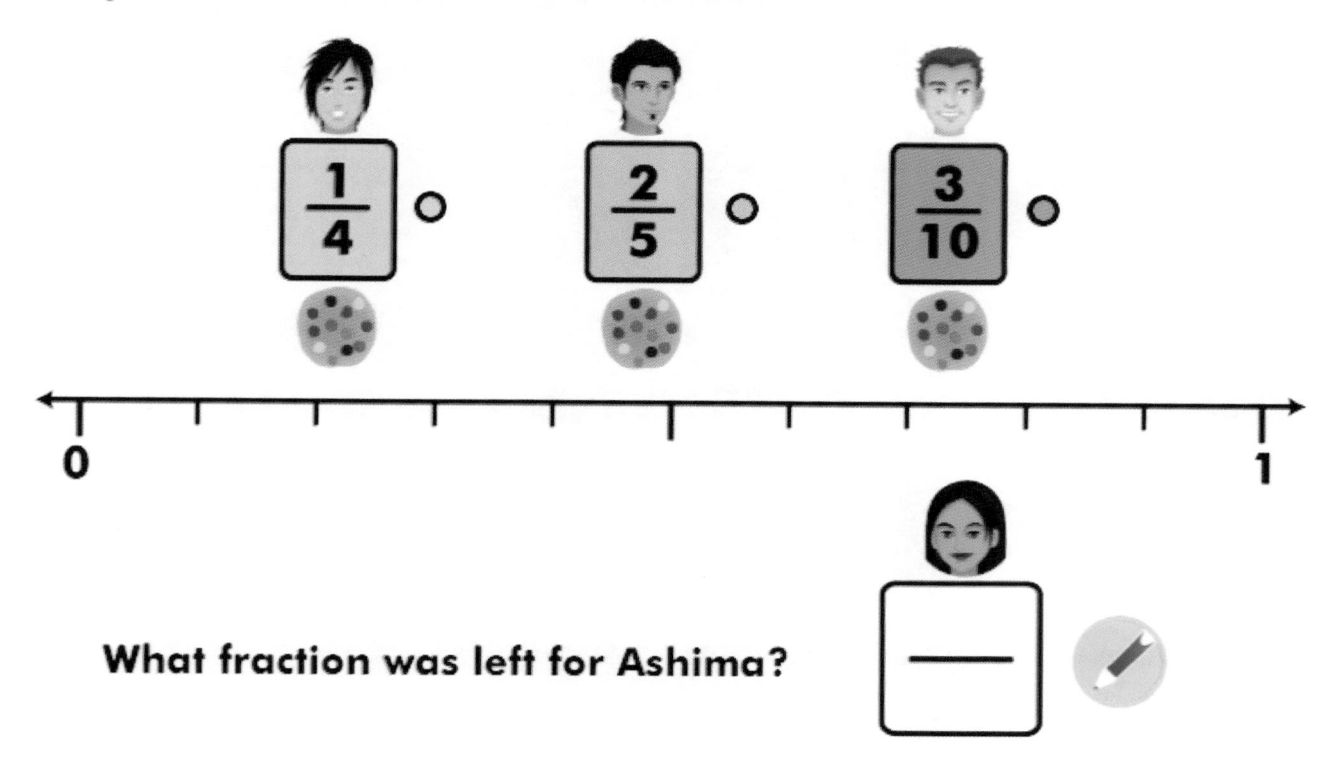

What fraction was left for Ashima?

Here's an exercise to stretch your new knowledge.

Using the number line to help figure out the problem, write each fraction as a decimal. Place a point on the number line and estimate the decimal.

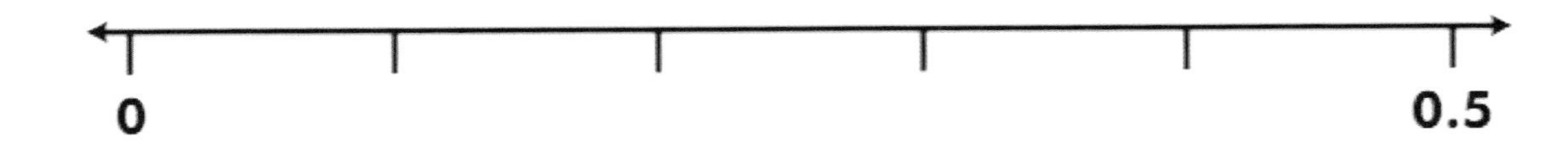

Compare fractions using the number line.
Draw a line from the correct point on the number line to the fraction.

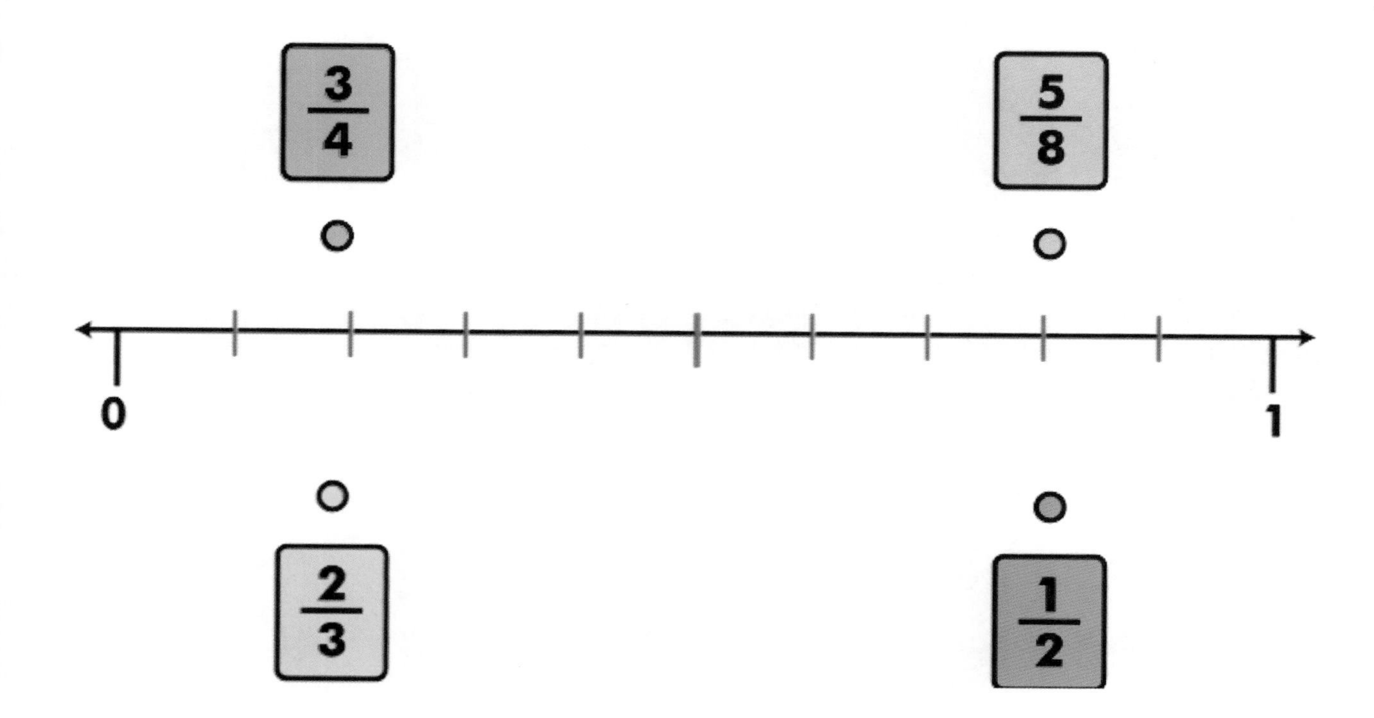

Name________________________________

Compare & Order Fractions Quiz

1. True or false: $\frac{7}{8}$ is smaller than $\frac{5}{6}$?

2. Which fractions are ordered correctly (small to large):

   A. $\frac{5}{8}$ $\frac{3}{5}$ $\frac{5}{7}$ $\frac{2}{3}$

   B. $\frac{3}{5}$ $\frac{5}{8}$ $\frac{5}{7}$ $\frac{2}{3}$

   C. $\frac{3}{5}$ $\frac{5}{8}$ $\frac{2}{3}$ $\frac{5}{7}$

   D. $\frac{3}{5}$ $\frac{2}{3}$ $\frac{5}{8}$ $\frac{5}{7}$

3. What is the LCD of $\frac{3}{14}$ and $\frac{1}{10}$?

4. Write $\frac{3}{8}$ as a decimal.

Newburyport, MA 01950

1-800-596-3175

OnBoard Academics employs teachers to make lessons for teachers! We create and publish a wide range of aligned lessons in math, science and ELA for use on most EdTech devices including whiteboard, tablets, computers and pdfs for printing.

All of our lessons are aligned to the common core, the Next Generation Science Standards and all state standards.

If you like our products please visit our website for information on individual lessons, teachers licenses, building licenses, district licenses and subscriptions.

Thank you for using OnBoard Academic products.

Made in the USA
San Bernardino, CA
15 January 2014